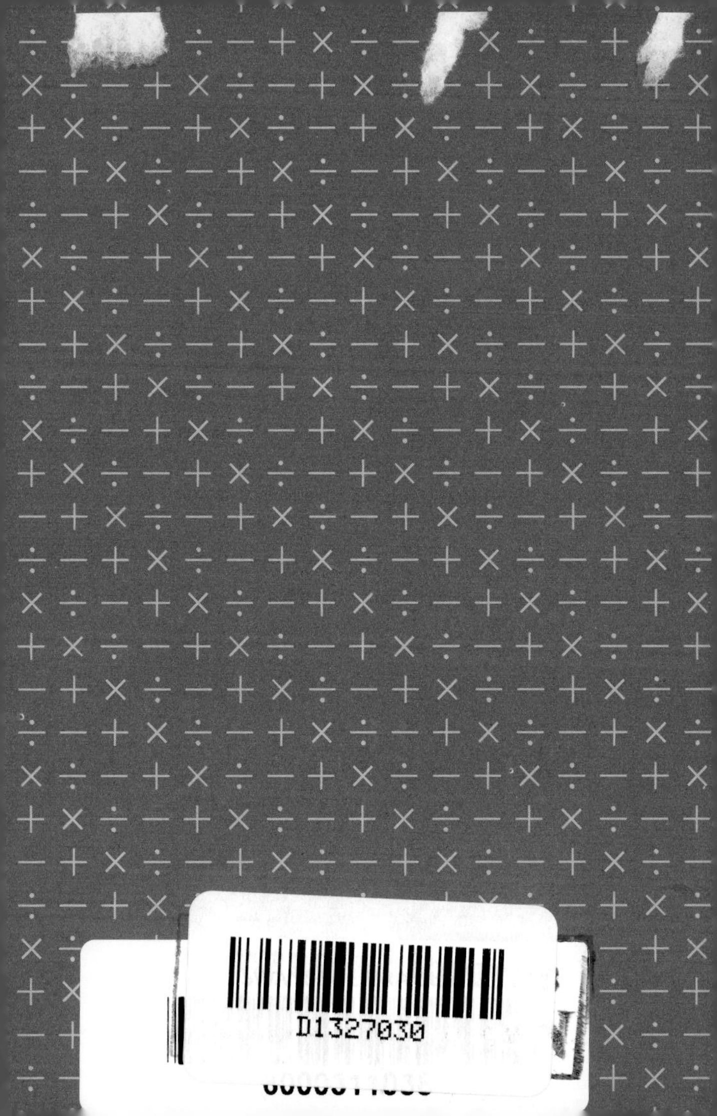
D1327030

SODDING SUMS

SODDING SUMS

THE 10%
OF MATHS YOU
ACTUALLY NEED

HYWEL CARVER

KYLE BOOKS

First published in Great Britain in 2017 by
Kyle Books, an imprint of Kyle Cathie Ltd
192–198 Vauxhall Bridge Road
London SW1V 1DX
general.enquiries@kylebooks.com
www.kylebooks.co.uk

10 9 8 7 6 5 4 3 2 1

ISBN 978 0 85783 445 4

Editor: Hannah Coughlin
Copy Editor: Caroline Taggart
Designer: Nicky Collings
Illustration: Aaron Blecha
Production: Lisa Pinnell

A Cataloguing in Publication record for this title is available from the British Library.

Printed and bound in Malta by Gutenberg Press Ltd.

CONTENTS

INTRODUCTION

You've probably realised it by now: 90% of the things your maths teacher said would come in handy one day have been totally useless. If you had £1 for every time you thought, 'I really wish I'd learned Pythagoras' theorem properly,' you'd probably be exactly £0 richer by now.

Unless you've started that PhD in Astrophysics, there's no reason for you to care about trigonometry, logarithms, the volume of a sphere or any of the other maths trivia your teachers tried to cram into your brain.

This book is about everything else that's essential.

There's a small amount of maths that *is* useful in everyday life, but which often gets skipped over or explained badly in school (in favour of yet more rules named after dead Greek men). That's what's in this book. Have you ever wondered whether you're getting a good deal in a supermarket? Or wanted to know how much a loan will cost you? Here are the answers.

I'm passionate about maths, and about making it accessible and useful. I'm going to explain some everyday situations where a little maths can go a long way, and I'll show my working, too, without getting too bogged down in all the sodding sums.

Everyone can benefit from understanding and using maths, so let's dive in.

ARITHMETRICKS

When you need to quickly work out which deal is best in the middle of the supermarket, or how many litres of beer to order, you can either do it in your head or use a calculator. Working things out on a calculator is more accurate, but mental arithmetic can be very fast (and is handy when you've left your calculator at home, or if you don't want to reach for one in a meeting).

You might already have a calculator application on your smartphone but, if you don't, it's worth installing one. Alternatively, online search engines like Google will do sums for you if you type them in. If you search for '5 + 6 x 3', it'll work out the answer.

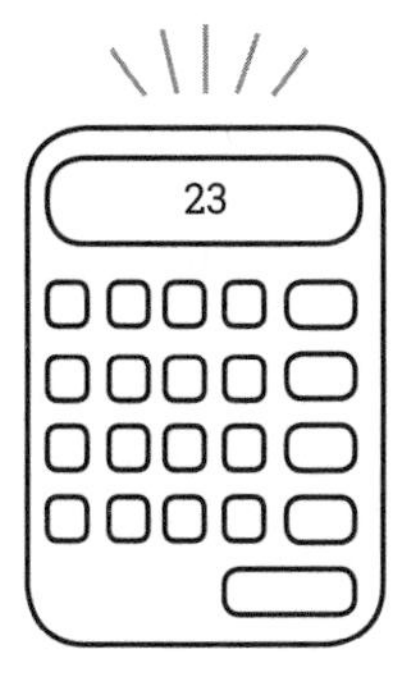

MATHS SYMBOLS

Here are some fundamentals that will help you understand all the sums in this book (or in case your kids want help with their homework).

Brackets – the ( ) that sometimes surround part of a calculation – mean you have to do that part of the sum first.

- (3 + 5) x 7 means that you should first do 3 + 5, which is 8, then use that result to do 8 x 7, which is 56.

If there are no brackets, **always** do division and multiplication before addition and subtraction.
To work out 3 + 5 x 4 ÷ 2 – 1 :

- Start with the division 4 ÷ 2 = 2 to get 3 + 5 x 2 – 1.

  Next do the multiplication 5 x 2 = 10 to be left with 3 + 10 – 1.

  Then the addition 3 + 10 = 13 gives you 13 – 1.

  When you do the subtraction you reach the answer: 12.

You can remember this order with the acronym **BODMAS**:

- first do **B**rackets
- then "**O**rders" (powers – see Loans on page 28)
- **D**ivision
- **M**ultiplication
- **A**ddition
- **S**ubtraction

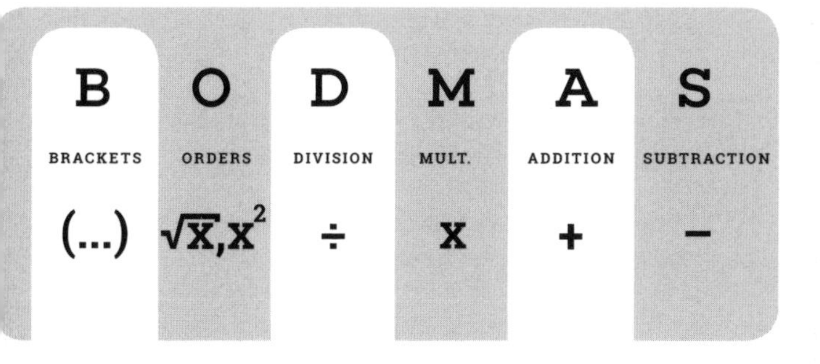

ROUNDING AND APPROXIMATION

Sometimes you just want to use maths to get a rough idea of something. Is this restaurant bill right? How many bags of flour do I need to buy for this recipe? How long will it take us to get there? Maths is a lot easier if we ignore some of the detail.

- 8059 x 2 is a hard sum, but 8000 x 2 is much easier and gets approximately the same answer.

Compared to the 8000 part of it, the 59 part is very small, so we can leave that out and get something approximately the same. This is called 'rounding'.

When we're approximating, we round our numbers by looking at their first two digits, ignoring zeros.

If the second digit is a four or less, round down. Leave the first digit alone and ignore all the other digits.

- To round 637, look at the first two digits: 6 and 3. Because 3 is four or less, ignore all the digits except 6 and get 600.

- To round a small number like 0.09215, look at the first two digits after the zeros at the start: 9 and 2. Because

2 is four or less, we ignore all digits except 9, and get 0.09 as our rounded version.

If the second digit is five or more, we round up – we still ignore all the digits except the first one, but make that digit bigger by one.

- To round 467, look at 4 and 6. Because 6 is more than 5, ignore all the digits except 4, make it one bigger and get 500.

This makes many sums easier.

- 19.73 x 5.12 can be approximated by 20 x 5 which is 100. The precise answer is 101.0176, so 100 is not far off. Most of the time, it's OK to approximate like this.

But how accurate are these approximate results in general? It all depends on the second digit. If the second digit is 0, 1, 8 or 9, then rounding doesn't change the number by much, so the results will be pretty close to the precise answer. If the second digit is anything else, rounding will make the answer you get less accurate, but will still give you a good idea of the figure you're looking for.

MULTIPLYING AND DIVIDING

When you're converting litres to pints or working out how much each item in a 3-for-2 offer costs, it's very useful to be able to do multiplication and division sums in your head.

DEALING WITH 10s

Multiplying by 10 is a neat trick – you just move the decimal point one space to the right. If there isn't a decimal point in the number, add a 0 onto the end.

- 10.3 x 10 = 103

  (moving the decimal point one place to the right)

  200 x 10 = 2000

  (an extra 0 on the end)

Dividing by 10 is the opposite: move the decimal point one space to the left. If there isn't a decimal point, put one in before the last digit.

- 29.3 ÷ 10 = 2.93

  309 ÷ 10 = 30.9

Simple!

MULTIPLYING AND DIVIDING BY 2

To double a number, add it to itself (you might need to round it first to save yourself some time).

- If tickets are £34 each, then 2 tickets will cost double.

  2 x £34 is the same as £34 + £34 = £68.

To halve a number, first round it to two digits – 78.9 becomes 79. Because doubling is much easier than halving, start by finding a number that will double to make 79:

- 35 doubles to 70, so we need something a little bigger.

  40 doubles to 80, which is too large, so less than 40.

  37 is roughly in the middle of 35 and 40 and it doubles to 74 (still not enough), so we need to go higher again.

  39 doubles to 78, which is very close.

We now know 39 is slightly too low and 40 is too large, so try the middle value of 39.5 – this does indeed double to 79 and so is our answer. Being able to double and halve in your head is useful for things like adjusting recipes for more people, and splitting bills in half.

MULTIPLYING AND DIVIDING BY 5

If you can do the tricks above, you can do this. To multiply something by 5, multiply it by 10 and then halve it.

- To do 29 x 5, start with 29 x 10 (by adding a zero to the end), which is 290.

  Halve it (as above: divide it by 2 by finding a number which doubles to make 290) to get 145.

To divide something by 5, first double the number and then divide it by 10.

- To do 29 ÷ 5, start with with 29 x 2 (by adding the number to itself) which is 58. Then divide it by 10 (by moving the decimal point one space right) to get 58 ÷ 10 = 5.8.

MULTIPLYING BY 9

There are two steps to this: multiply the number by 10, then subtract the original number from that result.

Let's say you want to buy 9 DVDs costing $14.95 each. Roughly how much will the total be?

The sum we need is:

- 9 x 14.95, which we can approximate as 9 x 15 (by rounding).
- Start by multiplying 15 by 10 to get 150, then subtract 15 (the cost of one DVD) from that to get 150 – 15 = $135.

Therefore, the 9 DVDs will cost a little under $135 in total.

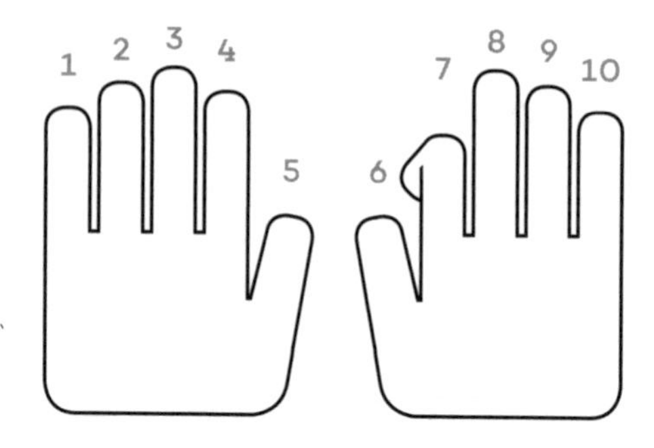

There's also an excellent trick for remembering your 9 times table up to 10 x 9. Hold out all ten fingers in front of you. Now bend the finger of the number you're multiplying by 9. So for 7 x 9, bend your seventh finger.

Count the fingers to the left of the bent finger – that's the first digit of the answer. Count the fingers to the right of the bent finger – that's the second digit of the answer.

- For 7 x 9, with your seventh finger bent, you'll have 6 fingers to the left of it and 3 to the right, giving 63 – which is what 7 x 9 is.

Teaching this trick to kids will blow their minds!

MULTIPLYING AND DIVIDING BY 3

To multiply by 3, add the number to itself, then add it on again.

- To find the cost of three crates of beer at €4.50 each, do 4.5 + 4.5, which is 9.

  Add 4.5 to that to get €13.50.

This might sound simple or obvious, but it's easy to forget when you're put on the spot and suddenly need to multiply something by three.

PERCENTAGES

You'll often come across percentages at work and in the news: 'Unemployment is down 0.5% this year' or 'Profits were 3% lower than expected.' You'll also see them in supermarkets, with signs saying '20% off' or '10% extra free'. Percentages are pretty important, but they can also be confusing – so it's worth understanding them well.

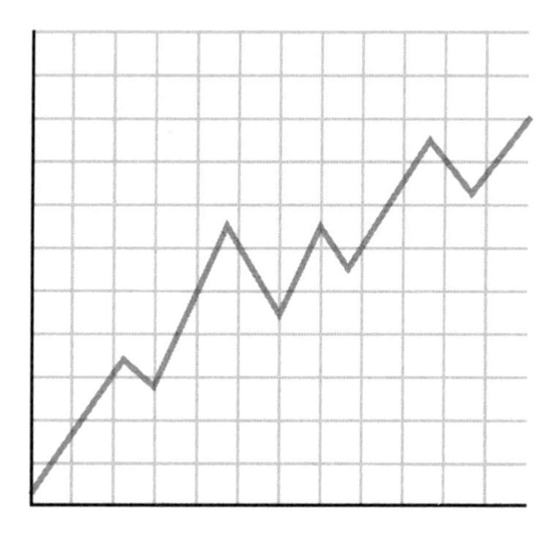

First, it's very handy to be able to convert between percentages and decimals. To turn a percentage into a decimal, simply divide it by 100. And to turn a decimal into a percentage, multiply it by 100.

- 'Susan drank 50% of a bottle of a wine' is the same as saying 'Susan got through 0.5 bottles of wine' because 50 divided by 100 gives 0.5.

If you have a fraction, you can convert it into a decimal by typing the fraction into your calculator, pressing '÷' instead of the '/' sign of the fraction:

- 'Susan drank ½ a bottle of wine.' Press 1, then the '÷' button, then 2, then '=', and it will tell you the answer is 0.5.

To find the percentage of something, convert the percentage to a decimal, then multiply them together.

- 19% of 500g = 0.19 x 500g = 95g.
  You may choose to do percentage sums on a calculator; 0.19 x 500 is a hard sum to do in your head.

One other important fact to remember is that 100% of something means all of it: no more, no less.

SUMS WITH PERCENTAGES

Let's start with adding a percentage to something.

- In a situation like 'You have to pay back the original €400 plus an extra 8%.' How much is that? An increase of 8% means that you have the original 100% plus another 8%, which is 108%. Using the 'divide by 100' rule, we can convert this into a decimal and get 1.08. Now we have something we can put into our calculator!

Multiply the decimal we found by the original number to get 1.08 x €400 = €432, which is the answer.

Decreasing by a percentage is similar, but now we have the original 100% *minus* something.

- Say your pension is worth 11% less than last year's total of $900. A decrease of 11% means you have the original 100% minus 11%, which is 100% – 11% = 89% of what you had before. Converting that to a decimal (remember: divide by 100) gives 0.89. As before, you multiply your original number by the decimal you found, to get 0.89 x $900 = $801.

How about this: 'UK house prices have increased 100% in the last 15 years and today the average house costs £230,000. How much did the average house cost 15 years ago?' In the previous examples, we had a number, did some percentage sums with it and found out what it would become. But things are more complicated when you are working backwards. Let's work through it.

- If house prices had started at £100,000 and then grown to £200,000, they would be 100% bigger (they're bigger by £100,000, which is 100% of what they started as). But the same problem working backwards would be: how much smaller were house price 15 years ago, compared to today? Well, they were £100,000 smaller 15 years ago, which is half of the average price today (£200,000), which, as a percentage, is 50% smaller, not 100%. The lesson here is that percentages work in one direction only: if prices grew 100% over the last 15 years, that doesn't mean they were 100% smaller 15 years ago.

If you knew the old average house price and were trying to work out the new average house price after an increase of 100%, you would do:

- 100% (the original price) + 100% (the extra price) = 200%, or 2.0 as a decimal.

Multiplying the old average house price by 2 gives the new price.

But because we have only the new house price, we're going to divide it by 2 instead to find the old price. In this example, that means that 15 years ago, the average house price was £230,000 ÷ 2 = £115,000. You might be able to do this in your head, but it's OK to do it on a calculator too..

It's the same when we're trying to work out what a number was before it was reduced by a percentage.

- 'The global population of tigers has decreased by 90.25% since 1950, to only 3900 left today.'[1] Because this is a percentage decrease, you first work out the decimal equivalent:

  You have 100% (the original amount) – 90.25% (the decrease) = 9.75% (what's left). Divide this by 100 to give the decimal 0.0975.

Because the population has already been reduced and we're working out what it used to be, we *divide* 3900 by this decimal to give 3900 ÷ 0.0975 = 40,000 which was the original tiger population in 1950.

1. These are real numbers! Gotta look after your tigers.

COMPARING WITH PERCENTAGES

How do you compare two offers fairly, for example if one brand of coffee is buy-two-get-one-free, and another has a 40% discount? The key thing is to do a division sum for each offer: divide the amount you have to pay by what you get for it (which must be numerical, like two coffees or 300ml). The answer with the smallest number is the best deal.[2] You might want to use a calculator to do these sums.

- If a cappuccino costs €2, the buy-two-get-one-free offer means you can get three cappuccinos for €4. To work out the cost of each cappuccino, use the 'amount you pay divided by what you get' rule:

  €4 for three cappuccinos means we calculate

  €4 ÷ 3 = €1.33 each.

  A 40% discount means each cappuccino costs 60% of €2 (because reducing by 40% means you're left with 100% – 40% = 60%), which is €2 x 0.6 = €1.20.

  Using the same 'amount you pay divided by what you get' division, you need to work out €1.20 ÷ 1 (because this is the price for one cappuccino) = €1.20.

2. Some supermarkets will print '16p a tablet' or '49p for 100g' on the price ticket, but if they haven't done that then it's up to you to work it out. The best value item is still the one with the smallest number, as it means paying less money per tablet or per 100g.

The best deal is the one with the smallest number, so the 40% discount is better (because €1.20 is smaller than €1.33).

Now, let's answer a crucial question. Is it better to have more gin for the same money, or the same amount of gin for less money? Gin 1 and 2 both normally cost £25, but today's offer lets you buy Gin 1 with 20% more gin, or Gin 2, 20% cheaper. And obviously you want to get the most gin for your money (if you're anything like me).

- **Gin 1** – pay the same amount, but get 20% more gin. 100% (one bottle) + 20% (the extra) = 120% of a normal bottle, which is 1.2 normal bottles as a decimal.

  Using the same division idea as before, find the cost of each bottle as the amount you pay divided by what you get: £25 ÷ 1.2 = £20.83.

- **Gin 2** – buy the same amount of gin, but pay 20% less. To find 20% off £25, as before, we do 100% – 20% = 80%, which (dividing by 100) is 0.8 as a decimal, then multiply 0.8 x £25 = £20.

  Use that division idea again: if you can get one bottle for £20, the division is £20 ÷ 1 = £20.

The rule is the smaller number is the best deal: £20 for Gin 2 is less than the £20.83 per bottle of Gin 1, so Gin 2 (20% off) is a better deal than Gin 1 (20% more gin). You can use the money you've saved to buy tonic water and a lemon.

LOANS

Loans are great for buying expensive things but can end up being very expensive themselves, so it's important to know what you're letting yourself in for. One concept you'll need to know which you might not remember from school is powers.

Powers are a way of writing several multiplications by the same number, written in superscript$^{\text{like this}}$. So the multiplication 9 x 9 x 9 is written as 9^3 (9 to the power of 3) because it's 9 multiplied by itself 3 times. And 4 x 4 x 4 x 4 x 4 is written as 4^5.

Powers are an important part of understanding how loans work. Most of the sums you do with loans are impossible to do in your head, so you'll probably need a calculator here. To do a power on a calculator, use the button marked ^. So 9^4 would be typed as 9 ^ 4.

For our purposes, a loan means that you are given a sum of money, and have to pay it back later with interest which gets added at a fixed monthly or annual rate. One of the reasons loans are so complicated is because of compound interest.

- You've borrowed $200 from Dodgy Dave for a new washing machine, and you'll pay it back in 12 months plus 10% monthly interest.

If you borrow $200 and the amount you owe increases by 10% each month (which is the same as saying that the interest is 10% per month), how much will you owe after 12 months?

You might think you have to pay back an extra $240. Intuitively, a 10% increase each month sounds like an extra $20 each month (because 10% of $200 is $20), so in 12 months that'll be an extra 12 x $20 = $240. With the original $200, that gives a total of $440. But that's not right.

The problem is that you pay interest on the interest (compound interest).

- After 1 month you'll owe 10% more:

  $200 x 1.1 = $220
  (see the previous section on Percentages, page 20, to understand why).

  After 2 months you'll owe 10% more than $220:

  $220 x 1.1 = $242

This is $22 more than the month before (*not* $20 more). After another month, the increase due to interest is $24.2 and the total you owe will be $242 x 1.1 = $266.20. In fact, after 12 months, compound interest means you have to pay back $628, not the $440 you might have thought.

Big difference! Here's how to work that out. First, convert the interest percentage to a decimal as you did when adding a percentage.

- 10% monthly interest means that until you pay back the loan you will owe an extra 10% each month, which translates to a decimal of 1.1 (100% + 10% divided by 100, as we did before). This means that each month the amount owed gets multiplied by 1.1.

Now we need to calculate that number to the power of the number of months (if the interest is calculated monthly) or years (if the interest is annual) you're going to have the loan for.[3] To do that, type in the decimal number you just found, then the ^ sign, then the number of months or years.[4]

- If your 10% monthly loan lasts 12 months then you need to find 1.1^{12}, which you type as 1.1 ^ 12, which your calculator tells you is 3.138.

For the last step, multiply the answer you just found by the amount you were going to borrow in the first place.

To buy the washing machine with Dodgy Dave's loan, that's 3.138 x 200 which is 627.6. So you'll be paying back $627.60 after twelve months for that $200 you borrowed. No wonder they call him Dodgy Dave.

---

3. The interest works by multiplying by 1.1 each month: after 1 month I owe £100 x 1.1, after 2 months it's what you owe after 1 month (which is £100 x 1.1) multiplied by 1.1 again, which is £100 x 1.1 x 1.1. So after 3 months it's £100 x 1.1 x 1.1 x 1.1. Multiplying the same number by itself is exactly what powers are useful for: the amount owed after 3 months is the same as £100 x 1.13. That's how the maths around loans works!

4. If your loan isn't for a whole number of years, calculate how many months you'll have the loan for, and divide that number by 12 to find the number of years.

Because of the compound interest of a loan it's easy to be surprised by how much money you will owe when you have to pay it back – it can quickly become double the amount you borrowed! Make sure you understand the total before you accept a loan.

Mortgages are similar to loans, except that you regularly pay back part of the money you borrow, rather than handing over one lump sum at the end. The maths on mortgages is too hard to fit in here, so I'd recommend using an online mortgage calculator instead.

Loans and mortgages are complicated. Check your sums with your lender and an independent advisor you trust before signing anything!

PROBABILITY AND GAMBLING

Games and gambling are some of the most interesting uses for maths in 'real life'. Understanding probability can make or save you lots of money.

Probabilities are numbers between 0 and 1 describing how likely something is to happen (an 'event'), where 0 means an event is certain to not happen, and 1 means it will definitely happen.

For example, 'The probability of Beyonce agreeing to sing at your wedding is 0.000001' means that if you ask Queen Bey to sing at your wedding, there is a one in a million chance of her agreeing to be there when you put a ring on it.

To find the probability of an event, you first need to find the number of all equally likely outcomes, then count the number of them where your event actually happens. The probability of the event is the second number divided by the first.

- What's the probability of rolling a 5 or more on a normal die? Well, each time you roll, there are six equally likely outcomes (one for each number that could turn up). Two of those outcomes would make your event happen (rolling a 5 or a 6), so the probability is two out of six, which is 2 ÷ 6 = 0.33.

Here's a gambling example:

- Donald will let you pay $1 to play a simple game in which you roll a die for the chance to win $4. So if you roll a 6 on a normal 6-sided die[5] he will give you $4. If you roll a 5, he will give you your $1 back, and if you roll between 1 and 4, he'll just keep your $1. Should you play the game or is Donald giving you a bad deal?

In this situation, it helps to imagine playing the game six times – rolling each possible number once. When you roll a 1, 2, 3 or 4, you win nothing; when you roll a 5, you get back $1, and when you roll a 6, Donald gives you $4. But for each of those rolls, you pay $1, so in total, you'll pay $6 and make back $5, netting you a loss of $1. So if you average across the 6 rolls, you lose $1 ÷ 6 (about 17 cents) on each roll. That means Donald makes 17 cents on average every time you play. Don't do a deal with Donald!

5. As in "All the women roll dice and the woman whose die has the highest number chooses the first key from the bowl". It definitely looks unusual, but it's right.

You can also use probabilities to work out the average gain of a gamble, to measure how much you'll win (or lose) or average. To do that, you need to know the probability of each outcome. Then you multiple each outcome's gain or loss by its probability, and add together all the resulting numbers.

- In Donald's game above, there's a 1 ÷ 6 probability of getting your $1 back and a 1 ÷ 6 probability of gaining $4 (it's 1 ÷ 6 because each outcome only happens when you roll one of the six numbers on the die). That means your average win each time you play the game is 1 ÷ 6 x $1 + 1 ÷ 6 x $4 = $0.83, or 83 cents. But it costs $1 to play, to win 83 cents on average, so overall you're losing 17 cents each game ($1 – 83 cents) – the same as we found before.

BETTING ODDS

Probabilities can also be written as odds – for example 'There are 4-1 odds[6] that a photo of Vladimir will show him fighting a bear.' If the first number is much smaller than the second, it means that the event described is very likely. If the numbers in the odds are about the same size, the event is as likely to happen as not, and if the first number is much bigger, it means that the event is very unlikely to happen.

You can see this by converting the odds to probabilities. To do this, divide the second number by the sum of both numbers. So odds of 4-1 are a probability of 1 ÷ (1+4) = 1 ÷ 5. That means there's a one in five chance of it happening – not very likely but still possible.

Odds that you'd get from your local bookie are a bit different. The odds above are telling you how likely something is to happen, but the odds from a bookie's are telling you how much money you'd get from your bet if you won. In this context, odds of 3-2 are saying, 'If you win, you'll gain £3 for every £2 that you bet.' So each £2 you bet would get you back £3 (that you gained) + £2 (the original bet, which you are given back if you win) = £5.

6. That '-' isn't the same thing as a minus. Odds of 4-1 would normally be read as 'four to one'.

Let's say you bet on Ed to win a dancing competition at odds of 150-2. If you bet £2 and Ed wins the competition, you get paid back your stake of £2 plus £150 profit (think of the odds as saying '150 for every 2 you bet'). If you're really confident in Ed's ability to tango and decide to bet £100, you'll get £7,600 if he wins: your stake of £100 plus profit of £7,500 (your bet of £100 is like 50 bets of £2, so you win back 50 lots of £150).[7]

7. You can work this out as your initial bet multiplied by the sum of the numbers in the odds, then divided by the second number in the odds. So £100 at 150-2 gets you 100 x (150+2) ÷ 2 = 100 x 152 ÷ 2 = 7600.

STREAKS AND THE GAMBLER'S FALLACY

There are two places where our intuition really lets us down when it comes to probability.

The first is the idea of winning streaks and losing streaks. This is the feeling that 'you're on a roll' because you have recently won (or lost) so many times in a row that you'll carry on winning (or losing).

With any game of luck, that just isn't how it works. The dice, cards or roulette numbers are just as likely to come up with a win or a loss as they were before your 'streak'.

Even if you feel as if you're doing so well that you couldn't possibly start losing, your streak will carry on right up until it doesn't.

The opposite of this is what's called the Gambler's Fallacy (and there's a clue in the word 'fallacy' that this isn't true).

The Gambler's Fallacy is the idea that things that haven't happened for a while are 'due' to come up, and are more likely to happen soon.

The British National Lottery maintains a list of the 'most overdue numbers' – the numbers that haven't been drawn for the longest time. At the time of writing, the top of that list is number 57; the Gambler's Fallacy is that 57 is overdue and 'more likely' to come up soon, but that simply isn't the case. In the next draw, 57 will be just as likely to come up as any of the other numbers, as it has been on every previous draw.

That might sound obvious to you (but it also might not). A famous example is taken from the Monte Carlo Casino roulette wheel. In roulette, a ball spins around a wheel until it rests on a single number around the edge, roughly half of which are black and half of which are red. On one day in 1913, the ball fell on black for 25 spins in a row.

What would you bet was the colour the ball landed on next? People often think that red is more likely to come up because it's overdue – but, no matter what has come before, each spin of the wheel is just as likely to come up red or black. In fact, that day in the casino the next spin was black again – the ball fell on black a total 26 times in a row before falling on red.

STATISTICS

Former British Prime Minister Benjamin Disraeli is sometimes quoted as saying, 'There are lies, damned lies and statistics'[8], which is a neat way of saying that statistics can be used to bend the truth. With a better understanding of them, you can avoid being hoodwinked.

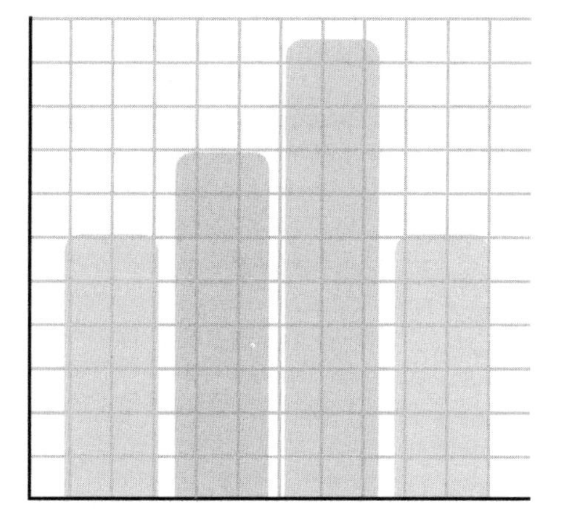

8. It's possible that he never actually said it, which is yet another demonstration that you can't believe everything you read (except in this book).

AVERAGES

Averages are a measure of the typical number of a group of values. For example, in the sentence 'His hands are much smaller than average', there's a comparison of one individual's hands against the typical or middle value from a group.

There's more than one way to measure that, though. Here are the two that you're likely to come across most often.

The **mean** is the form of average that most people use most of the time. (If the type of average isn't specified, it's probably a mean.) To find the mean, you add up all the values, then divide by the number of values in the list.

- The three Star Wars prequel films have average ratings of 6, 6.7 and 7.3 out of 10, so to find their mean rating we first add up all the values (6 + 6.7 + 7.3 = 20), then divide that total by the number of values there are (in this case, 3 ratings) which gives 20 ÷ 3 = 6.67 out of 10.

The **median** is the middle value in the group (remember this: it's a bit like 'medium' which is the middle value between 'low' and 'high'). To find the median, you have to put the numbers in order and look for the middle one (if there is an even number of values, find the mean of the two in the middle).

- Liz Taylor's seven husbands were 6, 20, 23, 4, 7, 5 and -20 years older than her respectively (the seventh husband was 20 years younger). In order from smallest to largest, they were -20, 4, 5, 6, 7, 20 and 23 years older than her, so the median husband was 6 years older (because 6 is in the middle of the list).

The median is useful when your data is clustered together but has a few outliers that aren't representative of the general trend. For example with Liz Taylor's husbands, the numbers -20, 20 and 23 are much more extreme than the rest, and could have a big effect on the mean.

CORRELATION AND CAUSATION

One of the ways statistics can be used to hoodwink you is by pretending that because two events tend to happen at the same time, one of them must cause the other. If two events tend to occur together, we say they are correlated.

Here's an example of two events being correlated but not causing each other. If you sat in your local park every day for a year, you'd notice that the percentage of people wearing shorts strongly correlated with the percentage of people eating ice cream. So you might conclude that wearing shorts *causes* people to eat ice cream. Of course, that isn't true, but not all examples are so clear cut. For instance, some studies have found a correlation between how much red wine you drink and reduced chances of suffering from cancer. So does drinking red wine reduce the risk of cancer? How can you tell?

You should first consider **confounding factors** and **controls**. Confounding factors are other factors that should be measured and considered as potential causes. In our question of whether wearing shorts causes people to eat ice cream, some other factors to measure would be age (maybe younger people wear more shorts and eat more ice creams) and, crucially, weather (hotter days might make people wear shorts and eat ice creams). Good studies measure confounding factors and consider whether they may be the true causes of the observed events; this is often referred to as controlling for those factors.

Real-world examples require lots of controls – for instance, to decide whether or not red wine reduces the risk of cancer, you'd have to control for diet (maybe some foods commonly eaten with red wine reduce cancer risk), lifestyle or wealth (maybe red wine is popular with certain social groups who also have hobbies or habits that have an effect on cancer risk), health status (maybe people with different overall health levels drink more or less wine, and also have an increased or decreased risk of developing cancer), etc.

If there's an important confounding factor that hasn't been controlled for, then there might be no causation between the two events. So, the next time you hear a news item saying that students who are taught grammar at school get better results in their other subjects, think about all the other factors that might affect this (the type of school, the supportiveness of their parents, the country they live in, how academic their peers are, etc.) and ask yourself whether those confounding factors have been properly controlled for.

CONVERSIONS

All of us prefer one way of measuring things. Temperature will always be in Celsius for me – I couldn't tell you whether 40° Farenheit means you should pack thermal underwear or flip-flops.

Useful, then, to be able to convert quickly from the measurement systems you don't know to the ones you do. These are all pretty accurate conversions, but if you really need precision you can search online for things like '8 pints converted to litres' to find an exact answer.

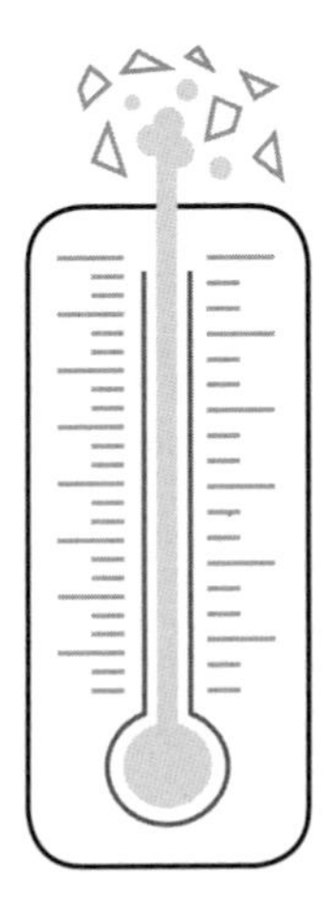

TEMPERATURE

To convert a temperature in Farenheit (F) to Celsius (C) exactly, subtract 32 from the value in Farenheit, then multiply by 5 and divide by 9. So $C = (F - 32) \times 5 \div 9$.[9] And if you want to convert a temperature in Celsius to Farenheit (e.g. to understand a weather report), multiply it by 9, divide by 5, then add 32 to the result, i.e. $F = (C \times 9 \div 5) + 32$.

If you don't have a calculator to hand (and you're not Einstein), you can get an accurate enough answer by subtracting 30 from the Farenheit temperature, then dividing by 2 to give Celsius. Or to convert Celsius to Farenheit, double it and then add 30.

$$C = (F - 30) \div 2$$

$$F = (C \times 2) + 30$$

Let's say you have a temperature of 40° Farenheit. To roughly work out the temperature in Celsius you do $(40 - 30) \div 2$. The first part, $40 - 30$, gives 10, so the answer is $10 \div 2 = 5$° Celsius. (The exact answer is 4.4°C, so we're pretty close.)

9. Remember BODMAS: work out the brackets first!

DISTANCE AND SPEED

The good news is that converting between distances is the same as converting between speeds – if you can convert kilometres to miles, the same sums convert kilometres per hour to miles per hour. Handy for driving abroad, especially if your phone contract doesn't include data roaming.

To convert kilometres to miles (or kilometres per hour to miles per hour), the precise way is to divide the kilometres by 1.6 to get the miles. And to convert miles to kilometres (or miles per hour to kilometres per hour), you multiply by 1.6. But with a slight simplification, we can make that much easier to do in your head and still get a pretty accurate answer.

To convert km to miles, first divide the number of km by 10, then multiply the answer by 2, then finally multiply by 3.[10]

$$\text{miles} = (\text{km} \div 10) \times 2 \times 3$$

- The Sun is 150 million km from Earth. In miles (and ignoring the million, which we'll put back in at the end) that would be (150 ÷ 10) x 2 x 3 = 15 x 2 x 3 = 30 x 3 = 90 million miles. The exact answer (150 ÷ 1.6) is 93.75, so again we're close enough for most purposes.

10. This is the same as multiplying by 6, but seeing as we've already seen tricks for multiplying by 2 and 3 in your head (see pages 15 and 19), I thought this would be easier.

To convert miles to kilometres, divide the number of miles by 10, separately divide them by 2, then add those two answers together, and add them to the original number of miles.

$$\textbf{km} = (\textbf{miles} \div 10) + (\textbf{miles} \div 2) + \textbf{miles}.$$

- 70 miles per hour in kilometres is $(70 \div 10) + (70 \div 2) + 70 = 7 + 35 + 70 = 42 + 70 = 112$ kilometres per hour.

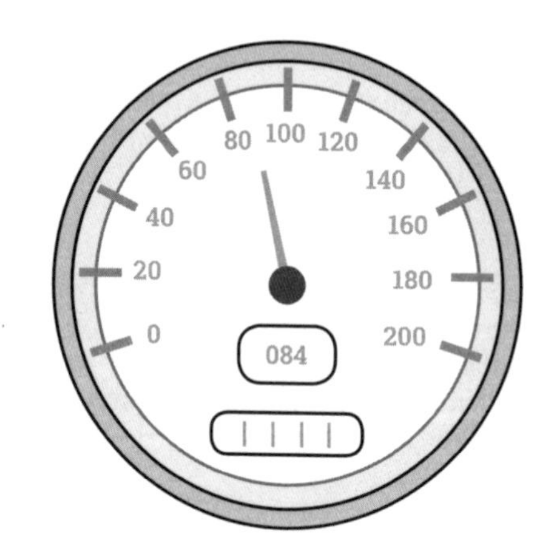

WEIGHT

To convert kilograms to stones, halve the number of kilograms, then add the answer to the original number of kilograms. Finally, divide the result by 10.

- With 80kg, do (80 ÷ 2), which is 40, add that to 80, which is 120, then divide by 10 to get 12 stone. This formula is approximate, but very close to the true answer (which is 12.6 stone).

stones = ((kg ÷ 2) + kg) ÷ 10

To convert stones to kilograms, multiply the number by 100 (like multiplying by 10 but add two zeroes on, or move the decimal point two places to the right), then halve it four times.

- 14 stone multiplied by 100 would be 1400, then halving it four times gives 700, 350, 175 and finally (approximately) 88 kilograms. Again, this is only roughly correct, but the exact answer is 88.9kg, so we're not far off.

kilograms = (stones x 100) ÷ 2 ÷ 2 ÷ 2 ÷ 2

To convert from kilograms to pounds, divide the number of kilograms by 10, then add the answer to the original number. Double the result and you have the answer in pounds.

- To convert 60kg to pounds, first divide 60 by 10 to get 6. Add that to 60 to get 66. Finally, double the answer to get 132 pounds.

$$\textbf{pounds} = ((\textbf{kg} \div 10) + \textbf{kg}) \times 2$$

To convert from pounds to kilograms, divide the number of pounds by 10, and subtract that from the original number of pounds. Then halve the answer to find the weight in kg.

- To convert 180 pounds into kilograms, divide 180 by 10 to get 18. Subtract that from 180 to get 162. Then halve 162 to find 81kg.

$$\textbf{kg} = (\textbf{pounds} - (\textbf{pounds} \div 10)) \div 2$$

To convert between grams and ounces, multiply the number of grams by 3 and separately halve it, then add those two numbers together. Divide the result by 100.

- You're halfway through a recipe which needs 400g of sugar, but your hands are covered in butter so you can't reach for your laptop to convert it to ounces. Well, using this conversion gives 400 x 3 = 1200 and 400 ÷ 2 = 200. Add those together to get 1400, then divide that by 100 to get 14 ounces.

$$\text{ounces} = ((g \times 3) + (g \div 2)) \div 100$$

To convert ounces to grams, multiply the number of ounces by 3 and then by 10.

- To convert 12 ounces to grams, do 12 x 3 which is 36, then multiply that by 10, which is 360g. This is approximately right: the exact answer is 340g.

$$\text{grams} = (\text{ounces} \times 3) \times 10$$

VOLUME

To convert pints into litres, divide the number of pints by 10 and add the result to the original number. Then halve that answer.

- 12 pints divided by 10 is 1.2. Adding that to the original number (12) gives 13.2. Dividing that by 2 gives 6.6 litres. Even in litres, your doctor will probably not approve of you drinking that much.

litres = (pints + pints ÷ 10) ÷ 2

To convert litres into pints, divide the number of litres by 10, then subtract the answer from the original number. Then double that.

- If a recipe needs 2 litres of water but your measuring jug only has pints, divide 2 by 10 to get 0.2. Subtract that from the original number (2) to get 1.8, then double that to get 3.6 pints of water.

pints = (litres – litres ÷ 10) x 2

If you need to understand gallons instead (e.g. for fuel), use these conversions:

gallons = pints ÷ 8

pints = gallons x 8

You can then convert the gallons to pints, then convert the pints to litres. If you need to convert between pints and gallons in your head then divide by 8 by dividing by 2 three times. To multiply by 8, multiply by 2 three times.

- To convert 30 gallons of fuel into litres: first convert the gallons to pints by doing 30 x 8 = 240 pints. Then convert from pints into litres by dividing your answer in pints by 10 to get 24 and adding that to the original number of pints (240) to get 264. Finally divide that by 2 to get 132 litres.

NUMBERS, GREAT AND SMALL

'The total global loss from the financial crisis of 2008 was 15 trillion dollars.'

Big numbers often come up in the news and in everyday life, such as when governments spend billions on new schemes or your email program won't send files bigger than 5 megabytes. But it's not always clear how big each number really is – what even is a terabyte? Is it like a pterosaur? Figures like 15 trillion dollars are too big to imagine, so they become meaningless and you can't really engage with the story. Maybe that's how they get away with it...

NUMBER	SHORTHAND NAME	PREFIX (AND SYMBOL)
1	None	None
1,000	Thousand	Kilo (k) e.g. in kilogram
1,000,000	Million	Mega (M) e.g. in megapixel
1,000,000,000	Billion	Giga (G) e.g. in gigawatt
1,000,000,000,000	Trillion	Tera (T) e.g. in teralitre (a trillion litres)

To help deal with big numbers, we use a system of shorthand names for amounts that get bigger by a factor of one thousand.[11] For example, a million is the same as a thousand thousands, and a billion is a thousand millions. At the same time, in the world of science, there are prefixes that signify these different big numbers. So a kilogram is a thousand grams, a megaton is a million tons, a gigabyte is a billion bytes. The table below summarises these numbers, including how much space that many marbles would fill (assuming each marble is about 1cm in diameter) and how much digital data is equivalent to that number of bytes.

THIS MANY MARBLES...	THIS NUMBER OF BYTES IS ENOUGH FOR...
...is one marble!	...a single character, like the letter A.
...would fill about a pint and a half – almost a litre.	...a small, low-quality image, or 1 second of low-quality audio.
... would fill 5 baths.	...a medium-quality image, or 1 minute of MP3 music.
... would fill a house.	...10 minutes of high-quality video or 300 photos on a modern smartphone.
... would half-fill the Burj Khalifa, the tallest building in the world.	...over 200 DVDs or 250,000 MP3 songs.

11. In the modern English-speaking world, that is. A previous British / European system of counting big numbers used factors of a million, so that a billion was a million millions and a trillion was a million billions (instead of a thousand billions). That system is very rarely used now.

INDEX

ACKNOWLEDGEMENTS

Thanks to Tom Carver and Charlie Reams for proofreading and checking my working.

Thanks to Hannah and all at Kyle Books for making this a fun book to write.

Thanks to my wife, Catharine, for helping me to write and proofreading this book, for her enthusiasm for the nine-times-table trick, and for generally being fantastic.